Encryption

[ ●]

for *babies*

by Jessica Dickinson Goodman

This is an idea.

This is your idea.

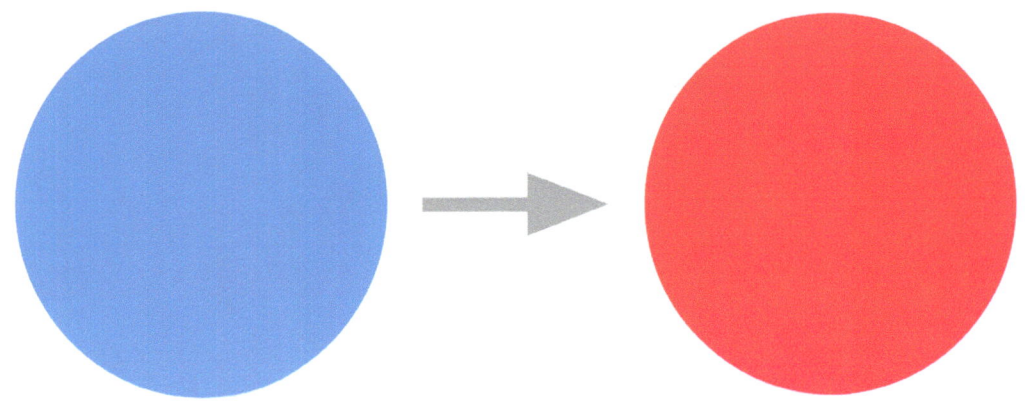

You get to choose who to tell about your idea.

This is encryption.

Encryption is a promise.

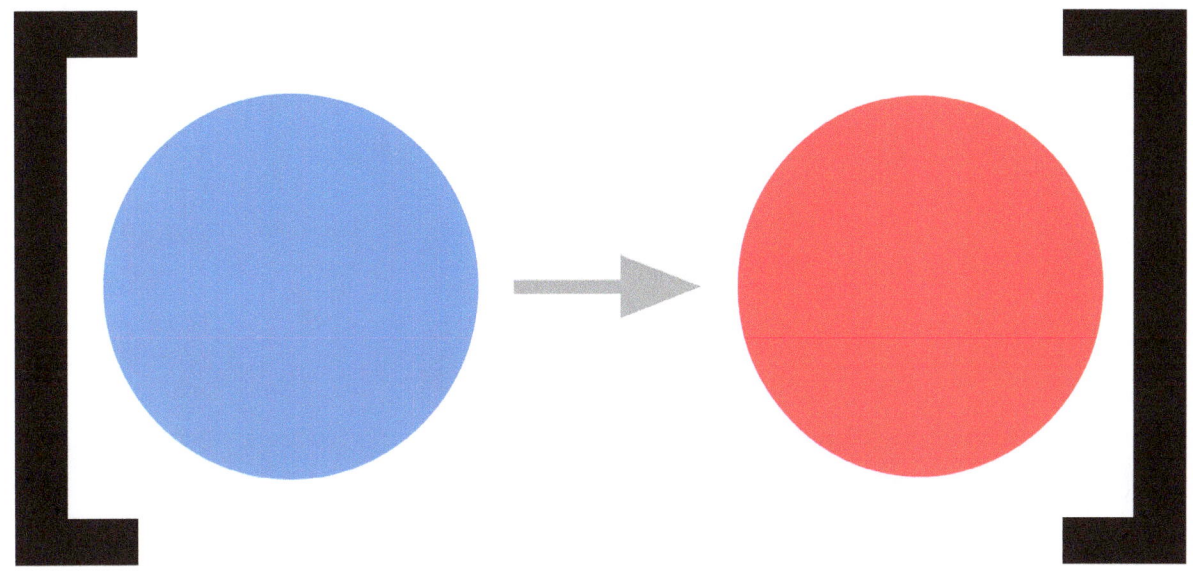

Encryption promises you get to choose who to tell about your idea.

This is a key.

Like a key to your door.

Encryption keeps its promises with keys. Your key unlocks your encrypted idea.

If someone wants to know your idea, they need to ask.

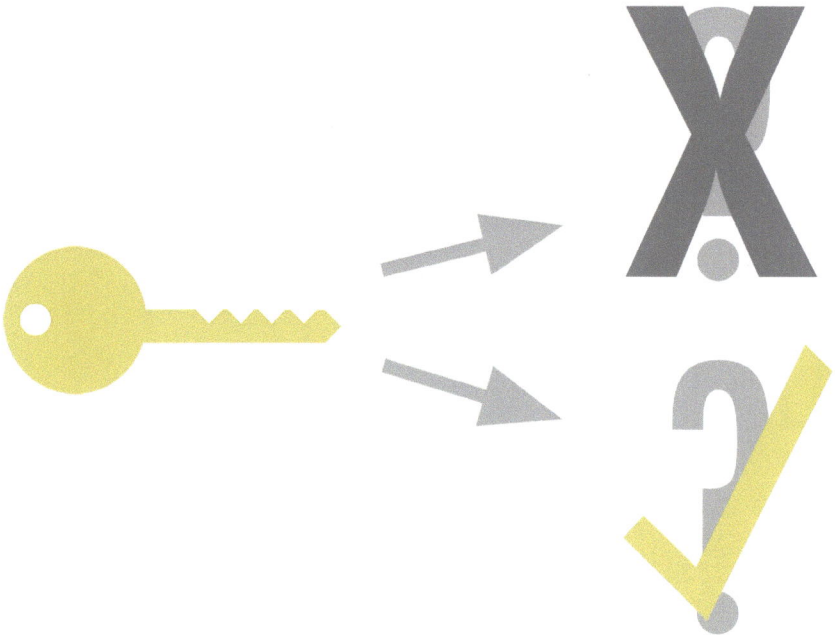

You get to choose who you share your keys with.

Some people want to know your ideas without asking.

But encryption is a promise.

And unless someone makes you break that promise,

**your ideas are safe
if they are encrypted.**

Some things are called encryption but don't necessarily keep their promises.

If someone else keeps a copy of your keys and can choose to give them away without asking you,

encryption can't keep
its promise.

You can trust encryption to keep your ideas safe if its called end-to-end encryption or e2ee.

[Now you know all about encryption!]

The Internet Society: San Francisco Bay Area Chapter

The San Francisco Bay Area Internet Society Chapter is one of the largest Internet Society chapters in the United States, serving the San Francisco Bay Area, Silicon Valley and the rest of California by promoting the core values of the Internet Society. Our goals are to: provide a platform for members to engage in Internet issues that include Internet governance, Data Protection, Privacy, Security, IoT, Internet Technologies, Standards and Access; promote the Internet Society's core values; interact with other ISOC Chapters and relevant Internet industry groups; disseminate information on the issues of importance to Chapter members. Learn more: sfbayisoc.org

The Internet Society

The Internet Society supports and promotes the development of the Internet as a global technical infrastructure, a resource to enrich people's lives, and a force for good in society. Our work aligns with our goals for the Internet to be open, globally connected, secure, and trustworthy. We seek collaboration with all who share these goals. Together, we focus on:

- Building and supporting the communities that make the Internet work
- Advancing the development and application of Internet infrastructure, technologies, and open standards, and
- Advocating for policy that is consistent with our view of the Internet

Learn more: internetsociety.org

www.ingramcontent.com/pod-product-compliance
Lightning Source LLC
Chambersburg PA
CBHW060822290526
45792CB00005BB/1761